TRUE

HAUNTINGS IN NEW ENGLAND

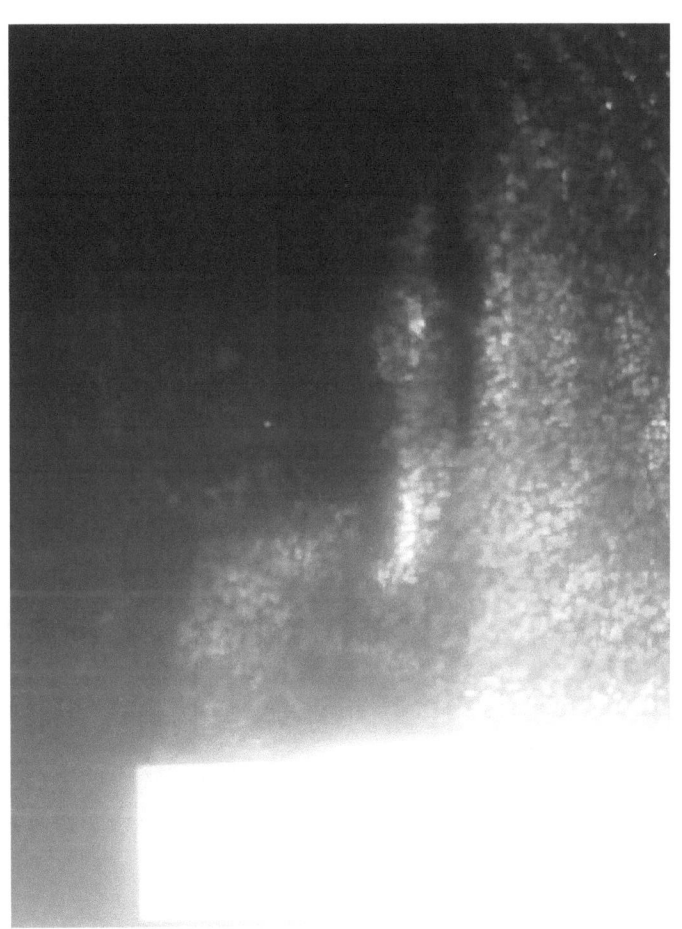

TOLD BY A PARANORMAL INVESTIGATOR
BY ERIC PERRY

FORWARD

First and foremost, My Mom Ann Perry, thank you mom for all your support. Barry Fitzgerald For all of your knowledge that I have enjoyed reading over the years.

To my former team Central NH Paranormal Society I have nothing but love for you all and respect.

To My new team I have learned so much with each of you. Each one of you taught me how to be grounded and shown people compassion and love for each of the clients. Haunted in New England will always be my Home.

To the Families and business that I have helped over the years without your support and believing in me and my team. It would have never happened and my journey into the paranormal.

And to all my Fellow Investigators; thank you for all your advice you Have Given me over the years.

Love you all ; but to one lady in my life Shelley THANK YOU BABYGIRL

The Beginning

It all started in Lynn Ma public school. Before you say anything it was during a field trip to Historic haunted Salem Ma. During this trip we House of the Seven Gables in Salem Ma.

I was with a group of kids and I decided to stay behind and check things out. I was very curious about things and I wanted to stay behind and look and touch things in this room.

Which was a huge mistake on my part .This will be the first time I see Shadow Man. When I started touching the artifacts, I saw this large black shadow of a man out the corner of my eye. Could it have been a trick of light or was it my mind playing tricks on me, what happened next changed my mind about it.

As I was touching the artifacts and I felt a cold breeze come by me and all of sudden I felt a touch on my shoulder. Ok at this point I ran like a little school girl.

My teacher got on the bus and asked me what was wrong? I wouldn't tell her anything.

Today as an investigator that would have been one of the best places to investigate.

Author Eric Perry
Author Eric Perry

Pine Grove Cemetery

Lynn, Massachusetts

The cemetery consists of approximately 250 acres of which 82 developed and houses between 88,000 and 90,000 interments. The cemetery has been recorded in Ripley's Believe It or Not as the "second longest contiguous stone wall in the world", second only to the Great Wall of China. The wall is built of field-stone and was built by the WPA in the 1930's.

Amongst the cemetery properties are architecturally distinct structures including the chapel, tomb and main office. Pine Grove Cemetery remains intact as a well-preserved example of early picturesque cemetery design.

The Rhodes Memorial Chapel, built in 1891, is among the landmarks of Pine Grove Cemetery and an innovative design of the late 19th century Richardson Romanesque style in Lynn. The donation by Mrs. Amos Rhodes, as a gift in memory of her husband, was generously granted to the City of Lynn. Over the years, it has been refurbished and the stained glass windows are valued at over $10,000. Pine Grove's name comes from the wealth of pine trees that frame the Rhodes Chapel and the entrance drive off of Boston Street.

The Pine Grove Cemetery Receiving Tomb was originally constructed during 1866-1868, part of a building campaign following the civil War, to modernize the facilities. The Tomb remains intact as part of the mid 19th century landscape design, made of granite ashlar construction with cast iron doorway in Ruskinian Gothic style.

The Cemetery Office Building on the grounds was erected in 1860, inside the Boston Street gates at the bottom of the hill leading into the Rhodes Chapel. The Pine Grove Cemetery gates, of cast iron are valuable examples of mid 19th century civic design, well incorporated into the landscaping of the cemetery.

The Cemetery's Greenhouse operations have been widely recognized across our country for the highly popular flourishing of landscape designs. The greenhouse has twin bays where flowers and plants are grown from seed. The plants are used for endowed lots and for flowerbeds in different locations around the cemetery and the greenhouse has a heating system to maintain temperatures year round.

But you want to know the haunting of Pine Grove Cemetery. As a teenager I worked at pine grove as a groundskeeper with a youth group for the city of Lynn Ma. There have been many murders and unexplained phenomena as well. It was during this time I came across what I believe to be a shadow person. I was

mowing the lawn on the forest rock side of pine grove. It was about 1 in the afternoon and it was a clear blue sky. I saw this shadow move from the left to the right ;and I could see right through it. Later that day it got even closer to me. It was 80 degrees out but it felt like it was 40 degrees out. That afternoon I felt I was being targeted by this spirit. As it came closer to me i felt so sad and scared of what i was seeing. Today I went back a few times with my team and just friends. And to this day I am nervous to go to the cemetery during that time of day.

As a haunted location you need to get permission to investigate it by the City of Lynn Ma.

This place has a lot of great history and is a paranormal hot spot.

Remember to be respectful of the dead

Haunting of High Rock Tower

Lynn, Ma

Lynn's High Rock is an outcropping of porphyry (crystal filled granite) a half a mile from the Atlantic Ocean, some 200 feet above sea level, the highest elevation around. The 85-foot high tower and observatory at the top of High Rock can be easily seen from Lynn City Hall and the downtown area. The poetry of High Rock is concerned with the view from High Rock and its elevation. Because High Rock is as close to heaven as a Lynn poet can get, it inspires divine thoughts in poets who visit its summit.

To get to High Rock, follow Essex Street away from downtown Lynn, and turn left up Rockaway Street. Then turn up High Rock Street and proceed to Circuit Avenue. Park at the end of Circuit Avenue where High Rock Park, the tower, and the observatory are located.

Hutchinson Family Singers Before English Settlement, High Rock was known to be a Pawtucket Indian meeting place, and a great chief named Nanepashemet once made his headquarters there. The modern history of High Rock is the story of its association with the Hutchinson Family Singers, who were for several decades America's favorite singing family. Family leader Jesse Hutchinson bought land at High Rock, and by 1846, he had erected the famous

Stone Cottage for his family. Other cottages were added, as well as a tower at the summit. During the Civil War, the family held concerts at High Rock. The Hutchinson's were committed to

progressive causes, and these included abolition, women's rights, and worker's rights. This commitment is expressed in the lyrics to the family's song, "High Rock": In the State of Massachusetts, In the good old town of Lynn, There's a famous range of ledges, As eye hath ever seen;Two hundred feet the highest point Looms up this rugged block, And it's known throughout New England As "Old High Rock."

And here the tribe of Jesse Sang and made the people talk Of the friends of right and progress, At "Old High Rock."

In 1904, in response to a civic movement to acquire High Rock for the city, aging family leader John W. Hutchinson gave the family land,

including the summit, to Lynn. In return, the city agreed to build a tower to replace the first tower, which had accidentally burned during a celebration of the Civil War's end in 1865. The tower and its astronomical observatory were completed in 1906. Since then, periodic improvements have led to the recently completed total renovation of the observatory and its telescope by Lynn Community Development. Visit the city of Lynn's website for more information about observatory activities and the history of High Rock.

The Hutchinson's family. Reports of one of the Hutchinson are still haunting high rock tower. If you ask some people who checks out the grounds at high rock you just might see the shadow man of stone cottage.

The inspirational nature of High Rock is well illustrated by one interesting moment in its history. In 1853, the Hutchinson family gave former Universalist minister and then obsessive spiritualist John Murray Spear permission to use their woodshed while he worked on constructing an electrical machine that would revolutionize life and be a New Messiah. Spear believed he could channel messages from dead spirits, and when the "Band of Electricizers," as he called them, spoke to him, they instructed him to build such a machine at High Rock. They also told him exactly how to build it. The crowning moment of the God Machine, as some came to call it, came in June 1854, when, with the machine completed, and about the size of a dishwasher, a pregnant woman, known only as New Mary, lay down in front of it while in labour. When she got up and touched the machine, Spear claims it moved, animated by what he called New Motive Power. Naturally, the claim proved controversial, but in retrospect, Spear's experiment at High Rock shows a pre-Civil War fascination with electricity and underscores the character of High Rock as a source of inspiration and spiritual energy.

Some of the legends of high rock also had witchcraft as well but not as well known as Salem Ma. Here a couple of the local witches. 1670? Ann BURT Lynn, MA.

1680 Margaret GIFFORD of Lynn, MA.

This is Ann Burt one of 3 witches of Lynn

These witches were less known: Moll Pitcher The most legendary High Rock resident was Moll Pitcher, a clairvoyant who lived on Essex Street at the base of High Rock for fifty years until she died in 1813. Many sailors whose fortunes she foretold spread word of Moll Pitcher up and down the New England coast. Moll Pitcher's legendary status was sealed when the great American poet John Greenleaf Whittier, a resident of Amesbury, published his first epic poem, "Moll Pitcher" in 1831. Moll Pitcher's cottage unfortunately, in the poem, Whittier is unkind to his subject. In Whittier's poem, Moll Pitcher incorrectly predicts a sailor's death, and upon his safe return, she is discredited and dies a miserable death. Later in his life, Whittier grew to dislike the poem, perhaps because of its unsympathetic portrayal of Moll Pitcher's appearance and her life. Consider these excerpts: She stood upon a bare tall crag This overlooked her rugged cot= A wasted, gray, and meagre hag;In featured evil as her lot. She had the crooked nose of a witch,And a crooked back and chin;

And in her gait she had a hitch; And in her hand she carried a switch, To aid her work of sin; A twig of wizard hazel, which Had grown beside a haunting ditch, Where a mother her nameless babe had thrown To the running water and merciless stone.....Even she, our own weird heroine, Sole Pythoness of ancient Lynn, Sleeps calmly where the living laid her; And the wide realm of sorcery, Left by its latest mistress free, Hath found no gray and skilled invader.

The house still stands on the northwesterly side of Essex street, nearly opposite Pearl street.

Directors Home

This is one of the most haunted places that I have ever investigated in NH. This place has left many scares on all of us. This place has a special place in my heart. This investigation took its toll on my investigators and teams. Central NH Paranormal Society was called in to do an investigation of this home that was left to decay to nothing. This building dated back to the French Indian wars. In June we started to do our investigation of this building and 4 others as well including a jail. When started we didn't have all the fancy gear that we have today. We had our cameras and recorders and some other gear. I didn't know what true haunting was until this place, and I mean it. During our 2 month investigation of this location. We got attacked, scratched, health issues. And as soon you enter the building they attack you. When myself and Sandy and one other investigator went in July and we got attacked right off the bat.

Many of the spirits are not out to do you any harm, but there were a few here that wanted to hurt you and get you to leave. During that evening my eyes turned red and I was swearing at the top of my lungs. Which is not like me at all? A friend took me outside to get fresh air, when I was outside I started to feel better.

We hadn't even finished our investigation that night. We decide to go back in and confront these spirits. Which was a big mistake on our part? When we got back in the place felt different to a lot of electric charged air around us. Our gear started to get drained and started to go dead. The spirits are coming to life. We had some encounters with some French soldier who pointed a gun at my forehead. After that encounter I left the building and sealed it.

This building is no longer standing and that is a shame.

This was one of the tunnels near the director's home. You can not access the tunnels at all anymore. Area is patrolled by Merrimack County Sheriff.

This case will remain with me forever!

Eagle Mountain House

This investigation was one of the investigations that tested my faith in god.

We set up this investigation in Feb. We need help with this case and brought 3 other teams to investigate.

The original version of this farmhouse inn opened in 1878. The place later expanded into two additional buildings to accommodate visitors.

When we started the investigation all the team arrived on time and split into teams and set out to investigate this

haunted location. What was in store for that night was something else?

I set out with my team heading to the third floor. We started investigating the attic of Eagle Mountain House. The reports in the attic were spirits of children running around on that floor. This was also an old employee's living quarters.

We set up a red ball in the main hall way to see if the kids would play with it. We had no luck with the ball. We left the ball and came back later to see if the ball had move, we had no luck. On the third floor we set up a laser grid and a camera. When we checked it they were shut off. We can't explain why, because nothing showed up on video. At 10pm we all meet in the lobby for a break, and to go over evidence. As we handed in our evidence. Some people decided to head to bed and call it a night. A half hour later we headed out again. My team of investigators headed into the cottage, which would be my biggest mistake.

I and one other investigator entered the building first. I felt this presence when I was on the main floor. As the rest of the team entered the building we started on the main floor of the cottage. We got a few things on the main floor but nothing concrete yet. Did I say yet? Because we started to hear loud bangs on the second floor. Before we headed up on the second floor I felt I was being pulled over to a corner of the building.

I started to walk over and one of the investigators pulled me back with the group. We heard a loud bang upstairs we started up the stairs, we didn't realize we got an evp (Electronic voice phenomena). When

reviewed the evp later on the recorder it said " get out" which we should have listened to before we went up stairs.

As we entered the second floor as I started to walk down the hallway I kept seeing this old man sitting in a chair and it started to concern me. Every room I passed i saw him there. Was I being target by this spirit?

As the team went to the end of the hallway, we noticed a bunch of boxes tipped over.

As we started to head back down the hallway I noticed the old man again and I felt a touch on the back of my neck. When my mom used to do this to me when I was a kid. This happened a few times. What I am not telling you is that a scrap booking club used an Ouija board a few weeks before this investigation. Half way down I felt I was being pulled into a room. I was standing in what I believe was a portal open by the Ouija board. I felt this hot breath coming down on top of me. I felt a presence with me and it was the old man.

I yelled for some help the dermatologist Joe Andrade came to my aid. He started to feel around and felt the air around me and said I was under attack by something. I believe it was the old man and he had an evil side. The team brought me back to the main house and sat me down on the couch; and asked everyone for help. They stood me in the middle of the room and 20 people laid hands on me. What was strange for me was i was having an out of body experience. I was seeing all his memories; how cruel he and his wife were to their kids. They used the belt.

They were not nice to their kids. The gentleman in this mark on me forever. To this day I haven't forgotten about him. His memory in my mind;but this also makes me a beacon in the spirit world. When I went home the next day he wanted to make sure I hadn't forgotten him at all : I WILL NEVER RETURN HERE.

This is the main house and carriage house.

The Bancroft Building

This building has a special place in my heart and always will. I have been in this building at least once. Outside more than I can count. I was granted access by a maintenance worker on the state hospital grounds.

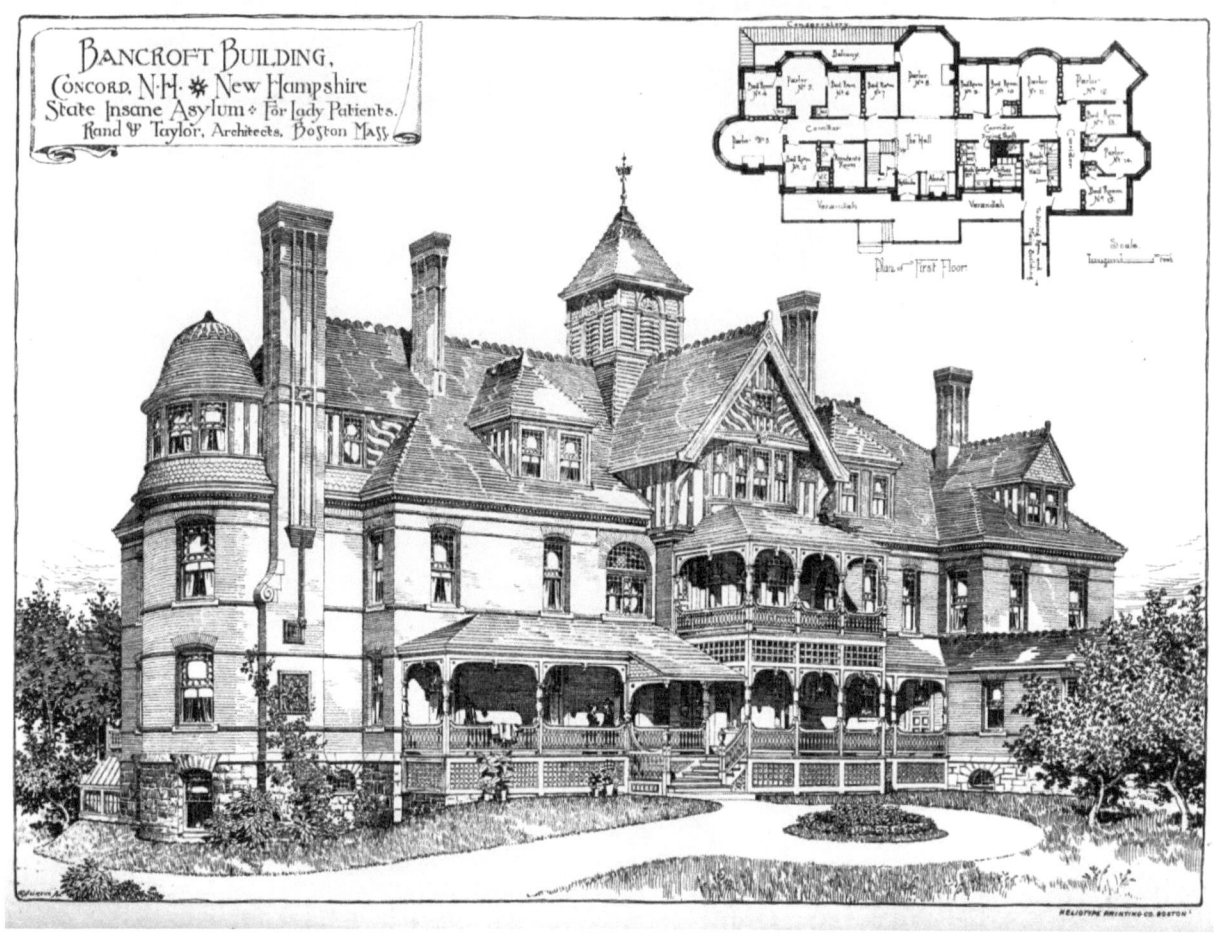

Picture from Concord Historical Society

Built in 1892 Bancroft provided additional space on the female side of the asylum and in part provided more ample, comfortable and private accommodations than previously available. Bancroft Building was modeled after buildings in Europe, which provided a more homelike, domestic appearance. It was a June afternoon and I decided to go and take pictures of this old asylum. I was walking around and this maintenance worker saw my CNHPS on and started to talk to me about some of the weird things going on in the buildings on the grounds.

I wanted to see more of the inside of these buildings and he said he could let me in a couple. He told me to hold on and said OK he took me to the 1920's1930's Hospital Residences buildings, which blew my mind.

Built in the 1920's-1930 these residences were originally designed for Hospital directors and medical care providers as an added stipend for their services, as well as to ensure their availability when needed. The Hospital also provided food, laundry, and light housekeeping services. The interiors of the residences were remodeled in 1989 to meet current standards for patient occupancy, and served as transitional homes for patients that were nearing discharge from the Hospital. Each home served as a vocational environment for patients to assist them in regaining essential skills they needed before leaving the Hospital. In 2012, Transitional Housing Services moved under the management of the Bureau of Behavioural Health. Then he took me to the 1880 Spring house.

Supplying water to an institution the size of the New Hampshire State Hospital was not an easy task. Shortly after the Hospital was built in 1842 a dam was constructed across Bow Creek to create a six-acre pond on grounds. Water was then pumped to a tank in the main building attic some 90 feet above ground. This supply of water was found to be inadequate to serve the growing Hospital population. The remedy was to sink a well in close proximity to the spring to collect water. The well, built in 1880, was 50 feet in diameter, 15 feet deep and provided 60,000 gallons of water daily.

All this history was amazing to see in person. Ok then i saw 1960 Anna Philbrook Center. A rumour of a nurse dying in this building was sad. The story goes like this, a nurse went in to check on a young man who was staying for a while. Wasn't taking his meds and choked out the nurse. Now do I believe these stories? No I do not. I cannot find any records of a death.

Does this mean they covered it up maybe.1960.

The Anna Philbrook Center is named for Dr. Anna Philbrook who was instrumental in its creation. Dr. Philbrook was one of the first women licensed to practice child psychiatry in this country and in 1933, the 63rd person to become a licensed psychiatrist in the United States. She started her career as a staff Psychiatrist at New Hampshire Hospital. In 1945, she was named director of the State Child Guidance Clinics. This unit had two homes the home of Ambassador John Winant and the basement of Thayer Building before a new facility was built on Hospital grounds in 1959. Until 1969, the North Wing of the Child Guidance This unit had two homes the home of Ambassador John Winant and the basement of Thayer Building before a new facility was built on Hospital grounds in 1959. Until 1969, the North Wing of the Child Guidance Center served as an outpatient treatment facility for emotionally disturbed children. In 1969, Dr. Philbrook's dream became a reality when the rest of the structure was completed as an inpatient unit and was named the Anna Philbrook Center. This building served children ages 4 – 15 until 2010 when the children's inpatient unit moved to the Acute Psychiatric Services building as consolidation of services occurred. Now let's get back to my favourite building. When I first saw The Bancroft , I loved the building and how it was built. But the paranormal investigator in me wanted to know what haunts this building and what secrets it has. Knowing that it housed the women of asylum. There were rumours of it also having children in this building with their mothers as well. Now was my chance to go in and find out if the ghosts of the past would show themselves. And one child spirit would let me know she was there. She was a ghost of days gone by. When you enter the building you notice all the beautiful wood work in this place. From the staircase to the floor was all hand done, and this building sits empty or is it? I was very excited to be in this building because I always just look at it from outside. I started on the first floor and worked my way up this building taking pictures all the way. I got to the second floor when I snapped a photo of what I believe to be a small girl standing near a window. She was a shadow of a little girl. It wasn't until I got home and started to go through my photos on the computer that I noticed her. Now I have been back to that building a few times. Even with a friend of mine in the NH STATE POLICE, Bill graham. Bill is a ghost hunter as well and loves to investigate the unknown. Ty Bill for telling me all your experiences you have had.

This was taken by I am Jen Photography

This was taken in the Bancroft Building on the 2nd floor. This is the photo that I had taken at the Bancroft building. As you can see the shape of what I believe is a little girl. This was taken during the day around 4pm. Concord State hospital has many different ghost stories, But I have proved that the ghosts of the old asylum are indeed haunting the Bancroft. In the main building of the asylum the reports of lovers who killed each other also can still be heard to this day. The story goes that a jealous boyfriend of one of the female patients thought that she was cheating on him with a male patient. He decided to bring a handgun into the building and he started yelling at her then it got heated and then he shot her, he dropped the gun she grabbed the gun and shot him. Today you can still hear the gunshots from these lovers. Concord state hospital doesn't allow paranormal investigation.

Valley Street Cemetery

Valley street cemetery caught me off guard. I did a 1 year investigation of this historic location. City records indicate that the Valley Cemetery land was donated to the City by Amoskeag Manufacturing Corporation in 1841. Through the Valley ran Cemetery Brook (known then as Mile Brook).

The receiving tomb was built in 1888, constructed into the hill north of the brook. In 1907, the gates at Auburn and Chestnut Streets were donated by Mrs Hannah A. Currier to honor her husband, the late Governor of New Hampshire, Moody Currier. The pedestrian gate, which was closed for many years, has been reopened.

The Gothic chapel was completed in 1932, replacing a wooden chapel that stood at the same site.

Deceased Manchester Mayors buried in "The Valley": Jacob F. James, Warren L. Lane, Frederick Smyth (the only Manchester Mayor to ever be elected governor of NH), Alonzo Smith, David A. Bunton, Darwin J. Daniels, Joseph B. Clark, David B. Varney and William C. Clarke. Private Mausoleums There are 13 private mausoleums in "The Valley". Haunted from the cemetery Thoughts of mass graves, sickness, and war already give a bleak look to the actual history behind Valley Street Cemetery, but what about the murder. Yes that's right murder. Smyth mausoleum, home to

41

NH Governor Smyth, was the site of a ghastly execution in the late 1970's early 80's. A prostitute, who often used the cemetery to service gentlemen, went to the Smyth with a man one night. While there it is believed he raped her, stabbed her, then threw her off the edge of the mausoleum (see picture on right) it is believed by many that her spirit still haunts this area. That she still sits upon the ledge of the crypt to warn people away. You do feel the present of a spirit when you visit Governor Smyth tomb. When we did the investigation we did notice that the cemetery has very Erie feelings as well. We have gotten some class an evps that was a class A. During the investigation we notice a group of trees in the shape of a pentagram, yes I said a pentagram. We ran emf meters and were getting strong reading from a couple of meters. Now these trees surrounded a grave stone which left us scratching our heads and wondering why someone would do that. Isn't that your final resting place? Putting these trees in a shape of a pentagram showing a form of disrespect? I say yes. If you have a loved one who has passed away, I would say no, then why do it. I believe that this cemetery has what is called a ley line.

Governor Smyth tomb.

Ley lines /leɪ laɪns/ are supposed alignments of a number of places of geographical and historical interest, such as ancient monuments and megaliths, natural ridge-tops and waterfords. The phrase was coined in 1921 by the amateur archaeologist Alfred Watkins, in his books Early British Track-ways and The Old Straight Track. He sought to identify ancient track-ways in the British landscape. Watkins later developed theories that these alignments were created for ease of overland trekking by line-of-sight navigation during Neolithic times, and had persisted in the landscape over millennium.

As my investigations of the historic cemetery gave me some answers. I can't call it a true haunting with one evp.

Remember not all cemeteries are haunted. And please respect the laws of these final resting places. Do not go in after dark unless you have permission from the local authorities or the cemetery itself.

ReMeMber pLeaSe ReSpect tHe Dead

Three Chimneys Inn

Location Durham, NH

I have travelled all over New England and seen many different things. I can't say every place is haunted. With the next place that I investigated I can say it is not haunted in my eyes, but many people say it is. And that's their own opinion as well and I respect that. One of these places was in Durham NH, known as Three Chimneys Inn. CNHPS went to investigate the inn on a rainy evening. We all showed up around 7pm and started to break into teams. We couldn't turn lights out because the restaurant was still open. And evps were also going to be a challenge also. Because the restaurant was very busy, we did try our best until it closed. When it did close we still got a lot of noise from the road. We tried different devices, ghost meters and a motion sensor lights turned on. But if that is enough to say it is haunted? No. We set out to debunk some of the claims of this haunting. This heater was high emf when we ran our meter across it. There has been a lot of investigation. If you are in Durham NH stop by and see for yourself and see the spirits of this historic location. It's a great place to visit in a college town.

Haunted Victorian Mansion

When you visit Gardner MA check out one of my favourite hot spots for paranormal activity. When I started investigating the paranormal this was one of the places I wanted to visit and investigate. My first investigation of the Victorian mansion was on a guide tour with the team. We started on the first floor of this historic home. Edwin told everyone the history of the location. Then we started to go up the stairs, but before that he told us about a portal on the left of the staircase. When you stand in it you feel like you are being pulled backwards. It was a crazy feeling. I have been back to the same spot more than once. And I am never let down. As we walked up the stairs I felt like I was being followed up the stairs. I stopped on the first landing and I turned around and took a picture on the stairs. And what I caught sent a chill down my spine when I got home.

The haunting of the Victorian is as real as you can get. We ended up on the 2nd floor and went into the red room. And Edwin told us about a death in this room of a hooker. Now I have been here 4 times and I have gotten some great evidence from this room. I have gotten class b and c evps. As I look back I remember watching ghost hunters about the Victorian. I remember that they had activity in the widow peak. We also learned about Spontaneous human combustion.Which means: Spontaneous human combustion (SHC) describes reported cases of the burning of a living (or very recently deceased) human body without an apparent external source of ignition. In addition to reported cases, examples of SHC appear in literature and both types have been observed to share common characteristics regarding circumstances and remains of the victim.

Forensic investigations have attempted to analyse reported instances of SHC and have resulted in hypotheses regarding potential causes and mechanisms, including victim behaviour and habits, alcohol consumption, and proximity to potential sources of ignition, as well as the behaviour of fire consuming melted fats. Natural explanations, as well as unverified natural phenomena have been proposed to explain SHC reports. Now back to my story: As we walked up the stairs to the 3rd floor I felt very uneasy; so did the kids as well; so the kids went out to the car. As we started to hear the history something caught my eye peeking around the corner. I saw a small child playing a game with me like peek a boo. Then he was gone but later on i would hear him. As I went up to the widow peak I felt something brush by me.

This is what I call the red room where the hooker had died in, some of us have named this room the hooker room. My thoughts on this place is that it has strong paranormal activity at this location. When I went to the basement I was in awe when I heard my first voice from the beyond of the little boy I saw up on the 3rd floor. I went into the well room and I was by myself when i heard a voice

My final thoughts; I have been all over New England and seen all kinds of haunts and non haunts. I got into this field to know if there is life after death. When you check your local cemetery please respect the dead and do not trespass on private property. If you are looking to join a team or group to investigate make sure you do the research about the team or group to make sure you are the right fit.

Since I wrote this in 2013 I have investigated this location 12 times and it is truly a haunted location and a historical location. The new owners have put a lot of work into the location and hope to open it as a bed and breakfast.

NOW GO FIND THOSE GHOST HERE IN NEW ENGLAND!

PEACE LIGHT AND LOVE

Terminology

Akashic Records: Originally a Hindu concept of a vast, and ever increasing, psychic repository of every thought and emotion - human or otherwise - which has ever been, and into which some individuals seem able to tap.

Agent: A living person at the site of a haunting. Some human agents act only as witnesses to paranormal events while others are believed to be the method by which the hauntings occur. Some agents may cause phenomena to increase, while others may be the entire source of the activity. How this works is as yet unexplained.

Amulet: A symbol with magical significance, which is worn as a pendant or ring.

Angel: "Messenger of God," a celestial being, benevolent in nature and if visible, appearing in human form, and possessing miraculous abilities such as teleportation, healing powers and knowledge of future events. There have been accounts of angels aiding people in times of crisis throughout the ages, albeit with no real consistency to their 'modus operandi.

Anomaly: An occurrence or condition removed from ordinarily understood experience. Apparition: The projection or manifestation of a quasi-physical entity.

Apport: A physical object that can materialize and appear at will and can include coins, watches, jewelry and even food. They are often connected to spirits who interact with the living as the spirits cause items to appear and disappear in an effort to make themselves known.

Astral Travel: Belief or theory that a person's spiritual awareness can temporarily detach itself from the physical body, remaining connected by what is called the "silver cord," and experience things in other locations, time frames or dimensional planes. Some refer to this as "Astral Projection" or "Mind Projection."

Aura-world: A reflection of our own sphere of existence, composed of the electromagnetic emanations of physical matter, and probably influenced by thought and emotion. It is another dimensional plane proceeding from one in which we exist.

Automatic Writing: A method used by spirit mediums to obtain information from the next world. It is believed that spirits take control of the medium and cause them to write unconscious information on paper.

Banshee: A death omen or spirit that attaches itself to certain families.

Banishing: Formal, ceremonial, procedure affected to cast an invisible presence or influence out from an area. This term can refer either to a spiritual cleansing, or the closing of a magical rite, when the invoked powers are dismissed.

Bogey: A spirit that is particularly antagonistic toward humans, traveling alone or in groups to cause trouble.

Channeling: A modern method of spirit communication in which the spirits pass information directly to the medium, who then repeats the information for the listener.

Clairaudience: The experience of receiving paranormal information through auditory impressions, voices and whispers. Many psychics are said to receive information from the spirit world in this manner. It can also be used to describe voices and whispers heard in haunted

Location.

Cleansing (Psychic): A less ritualized form of exorcism, where-in a dwelling or site is purified and malevolent influences are banished through prayers, spoken as the petitioner moves through the area.

Collective Apparition: A type of ghost sighting that occurs when one or more people see the same apparition.

Control: A spirit who acts as a medium's connection with the next world. Also referred to as a "spirit guide".

Construct, Psychic: It has been theorized, and experimentation has been conducted to support this premise, that through directed psychic energies a responsive spirit-like entity can be created, continuing for a time to exist independently.

Crypto-zoology: The branch of paranormal research which deals with the exploration of legendary creatures such as Bigfoot, lake and sea monsters,

Thunderbirds, etc. It should be noted that the Giant Squid (the "Kraken"), orangutans (the "Red Men of the Forest"), Komodo Dragons and gigantic Nepalese elephants all were formerly included in the roster of fabled creatures!

Crystal Skulls: Five human skull models, exquisitely crafted in antiquity from solid quartz crystals have been found in various locations throughout Latin America, the best known of these being the 'Mitchell-Hedges Skull,' discovered in 1924 in the Belize Jungle of Labuton by Anna Mitchell-Hedges while on an expedition with her father, and still in her possession in Canada. The others are kept in collections in Guatemala, Texas, the Smithsonian and the British Museum. Mayan legend tells that eight more crystal skulls remain, and that by the time all thirteen are united, mankind will have learned how to extract and decipher the vital information, history and revelations, which they contain.

Demon: Hostile and resentful entity, supposedly of non-human origin, which some believe to be "fallen (from grace) angels."

Dematerialization: The sudden disappearance of a person or spirit in full view of witnesses.

Discarnate: A word used to describe a spirit or specter...literally means "without flesh".

Doppelganger: Meaning "double image", it is thought to be an exact spirit double of a living person. They are considered to be very negative in nature.

Earthbound: Refers to a ghost or spirit that is unable to cross over at the time of death. Many spirits make the decision to remain behind by choice while others are too confused or frightened because of a sudden death or suicide to make the crossing.

Ectoplasm: An organic material that was supposedly exuded by physical mediums during séances as a way of proving contact with the spirit world. It would often take certain shapes. The substance was supposed to appear from just about any orifice of the medium's body. In more recent times, many researchers believe the substance had a natural form, created by fraudulent mediums during the Spiritualist era.

Electronic Voice Phenomena (EVP): Voices and sounds that are alleged to be from the dead and that are captured by electronic mediums on tape (digital recorders, tape recorders, etc.). Disembodied "voices" and sounds imprinted on audio recording devices.

Elementals: In magical tradition and ceremony, spirits which govern the four corners of the earth and are associated with, or reside within, the four basic elements. They are called Sylphs (the east, air), Salamanders (the south, fire), Undines (the west, water), and Gnomes (the north, earth).

Empath: An individual who is particularly sensitive to the psychic emanations of his or her surroundings, even to a degree of telepathically receiving and experiencing the emotions of others in their proximity. Obviously, psychic empathy can be regarded as a mixed blessing, and the empath must learn to gain a measure of control over this ability.

Entity: A disembodied "consciousness" commonly referred to as ghost, spirit or (if of an apparently malicious or resentful nature) demon. Exorcism: Ceremonial expulsion of invading spiritual/demonic entities from a person or dwelling, present in virtually every worldly culture. The Jewish and Catholic Christian faiths each have a formal 'Rite of Exorcism' to be conducted by the respective Rabbi or Priest.

Extra: A shape or a face that is said to have supernaturally appeared on film and cannot be explained away as fraud, faulty film or developing flaws.

Extra-terrestrials: Life forms originating on planets other than our own. This term usually refers to highly advanced visitors from other worlds, who journey to our sphere in space crafts with the probable intention of observing and studying our species.

Fetish: Aside from the modern sexual connotation, a fetish is a shamanistic tool in the form of a figurine, animal part or a pouch containing items with magical associations.

Floating Orb: A spherical image, usually translucent white, though sometimes of a reddish or bluish hue, which inexplicably registers on photographic film and videotape, also known as "Globule."

Ghost: The image of a person witnessed after his/her death, reflecting the appearance of the living, physical body yet less substantial. These forms often seem to exist in a dream-like state of semi-awareness, at times though not always cognizant of their human observers. Also, a generic term used for a number of different supernatural entities.

Ghost Hunting: Various methods of investigating reports of ghosts and hauntings and determining their authenticity.

Ghost Lights: Strange balls of light that appear in specific locations, often for an extended period of time but which have no explanation. They are thought to be of natural origin, possibly pertaining to earthquakes, fault lines, railroad tracks or water sources, but remain a mystery. Most such lights have a legend attached to them, usually involving a person who has been beheaded. The light is then explained as this person searching for their missing head. Globule: An anomaly where-in floating, circular forms appear on photographs or videotape, which seem indicative of spirit activity. Globes are a natural containment formation of the meniscus of liquid, as in gas containing bubbles; perhaps the interaction of energy and a quasi-physical substance produced by spiritual manifestations results in a similar effect, the globules being an initial containment of energy. Presently, all we know is that they continue to appear, and extraneous possible causes such as moisture, light refraction or emulsion seepage, etc., have been considered and ruled out. Golden-rod: A rare anomaly seen in videotape recorded at the site of a suspected haunting, appearing as bright, white or yellowish lines rapidly moving across a room.

Haunting: The repeated manifestation of a supernatural phenomena attached to a specific locale. The activity may appear as physical apparitions, sights, sounds, smells or cold areas. Hauntings may continue for years or may only last for a brief period of time. Hauntings can be categorized into four (usually) distinct types, these being Intelligent (responsive), Poltergeist (likely initiated by pent-up stress on a subconscious level), Residual (replay) and Demonic (non-human origin). Hex: A magical working, or "spell," cast to influence a person's will or fate, most often referring to a curse rather than a blessing or healing.

Hobgoblin: Mischievous sprite (fairy, spirit) who delights in perpetrating pranks upon hapless humans, once widely believed in and dreaded throughout Europe and Celtic regions. (Caution: It is theorized that these diminutive denizens of the netherworld will, upon occasion, interfere in psychic investigations by devices such as misplacing directions and telephone numbers, draining flashlight and camera batteries, and even pulling keys right out of investigators' pockets!) I assume that anyone who reads the proceeding caution will realize it is farcical!

Hypnosis: A state of profound mental focus, actually self induced although an external agent - a "hypnotist" - often acts as the catalyst, or director, for the subject entering this state. Also known as "Mesmerism" after Franz Anton Mesmer who first popularized this practice (utilizing magnets as his props) during the last two decades of the 18th century. As concerns paranormal investigation, hypnosis is sometimes used as a vehicle for "past lives regression" and memory restoration in suspected (alien?) abduction cases. **Icon**: A rendering or image of particular (often religious) significance.

Incubus: Stemming from medieval lore, a demonic entity capable of sexually arousing and sometimes assaulting human females. Cases of apparent incubus attacks continue to be documented, suggesting a germ of reality behind the myth.

Infestation: Repeated and persistent paranormal phenomena, generally centered around a particular location or person(s). Also known as a haunting.

Influence: An invisible entity of undetermined nature, affecting the inhabitants of a dwelling. This may initially manifest as an inexplicable feeling of uneasiness, then be followed by more definite signs which reveal a haunting.

Lepke: A very unique and interesting type of spiritual manifestation; a ghost which has the appearance of a solid, living person; may even converse with someone, then suddenly vanishes: "We were talking; I turned to face her again, and she was just gone!" Such apparitions are most often reported to have been encountered within, or immediately outside of cemeteries.

Levitation: A phenomenon sometimes encountered in hauntings, particularly with Poltergeists, rare yet credibly reported, where solid objects (including persons) are moved and lifted by an unseen force. The first historically documented occurrence was that of St. Francis of Assisi in the 14th century.

Lurking Enigma: "Lurk" means to furtively move about, and I can think of no more appropriate term to describe this phenomenon – a type of entity which can be visible to human observers, yet appears in distorted, unidentifiable forms. Common traits reported by witnesses include glowing red or silver eyes, dark color (fur or feathers), startling speed and agility, in some cases winged and capable of flight, as with the 'Jersey Devil.' Although such nebulous creatures seem to mean us no harm, encounters with them can be terrifying, and provoke much curiosity. As one would expect, they are extremely elusive.

Miracle: A wondrous and beneficial event apparently brought about by a supernatural/divine agent.

Materialization: A ghost appearing visually, suddenly or gradually, sometimes indistinct, sometimes seemingly quite solid.

Matrixing: The natural tendency for the human mind to interpret sensory input, what is perceived visually, audibly or tactilely, as something familiar or more easily understood and accepted, in effect mentally "filling in the blanks."

Mumiai: Native American Indian spirit which behaves in the manner of a Poltergeist.

Necromancy: The practice of communicating with the dead to obtain knowledge of the future, others' secrets, etc. An archaic term, the necromancer was said to employ magic spells and conjuration to summon, then banish, the spirits of the dead. Oui-ja (Board): A piece of wood bearing the letters of the alphabet that is used as a tool to make contact with the spirit world. Sitters place their fingers lightly on the planchette (or pointer) by which the spirits can spell out messages on the board. Experienced researchers vehemently advise against their usage.

Pact: The belief, prevalent in the late middle ages through the Renaissance, that someone could trade his or her soul in return for worldly gain.

Paranormal: A word meaning "unknown" or "beyond the normal" that has come to refer to events that are unexplainable. Parapsychology: The avenue of paranormal studies and research relating chiefly to psychic abilities (e.s.p., telepathy) and spiritual phenomena.

Pentacle/Pentagram: The traditional five-pointed star design, with its interior pentagon delineated, generally representing both spirituality and protection when pointed "up"; when inverted, it is said to signify diabolism.

Phantom: Another name for "ghost" or "spirit" although, interestingly, many use the word "phantom" to refer to ghosts that have been seen wearing cloaks or robes.

Phantom Lights: Sometimes they can be attributed to blue methane flame produced by swamp gas, or electrical discharges in the form of what is termed ball lightning or perhaps even misplaced fireflies. Yet, in other instances, the phenomenon of floating lights observed over water, the edge of woods, , lonely back roads and in the windows of darkened houses just can't be dismissed by ordinary explanations. These might be globules which coalesce and intensify in luminosity to the point where they become visible in dark surroundings.

Poltergeist: Literally means "noisy ghost" in German.

Although it actually refers to Traditional ghosts and hauntings, in other cases, it can be used to describe the work of a human agent. In this situation, the knockings and the movement of objects is caused by an outward explosion of kinetic energy from the human mind. Most Poltergeist outbreaks are short lived.

Possession: Invasion of the human mind by a spiritual or demonic entity, where the invading agent for a span of time, influences or entirely subverts the personality of the human host. It is in these instances that the boundaries of psychology, religion and spiritualism are rendered less distinct.

Precognition: The psychic perception of future events or conditions.

Psychic: An all-encompassing word that is used to describe a person who is allegedly sensitive beyond the normal means. Such a person may be able to see and hear things that are not available to most people.

Psychic Vampire: This is a term for individuals who seem to instinctively draw and absorb the psychic energies from others, usually while conversing with (or at) them.

Psychokinesis (PK): The ability to move physical objects using only the power of the mind. In many poltergeist-like cases, human agents affect objects in an unconscious manner.

Radiant Child: The apparition of a child which is seen glowing or surrounded by a bright aura.

Rapping: Can refer to sounds that occur at a location that is experiencing a haunting or be one of the earliest forms of spirit communication in which

mediums and spirits work out a code by which questions can be asked and then answered by raps from the spirits.

Reciprocal Apparition: A rare type of ghost sighting when both the spirit and the human witness see and respond to one another. Reincarnation: The belief that a person's soul will, following bodily death, inhabit a new body in a long cycle of rebirths, purportedly for the soul's evolution through gaining experience.

Residual (Haunting): Psychic imprint of a scene which is repeatedly played out, where the witness of such phenomenon essentially is peering into the past. The ghostly participants of these time-displacements often seem unaware of their living observers.

Retro cognition: The psychic perception of past events or conditions.

Revenant: An entity which projects an appearance of being distressed or misplaced.

Sanguinor: A person exhibiting vampirism tendencies (the desire to ingest blood) and attributes. These may be either contrived or pathological.

Satan: Hebraic term for "Adversary," the "Tester" in the Biblical Book of Job, the most familiar name of the Devil, the

"Fallen Angel" and the "Evil One." Investigators sometimes come across evidence of the activities of satanic cults, who perform animal sacrifices and apparently believe that desecrations and obscenities are devotions to their dark lord.

Séance: A group effort to contact the spirit world. In standardized format, the lighting of the chamber in which the séance is conducted is subdued, and the participants sit around the table, either holding hands or with hands palm down, flat against the table's surface and with fingertips touching those of the adjacent partners. A candle generally is set on the center of the table. The appointed director or "medium" addresses the spirit(s) with whom contact is sought, and then it's "We await a sign..."

Shade: An entity resembling a once-living being (human or animal).

Shaman: A tribal priest who, following much preparation and rite of initiation, uses the forces of magic to affect healings and divinations.

Silky: A female ghost which is attired in a rustling silk garment (sometimes seen, other times just heard) and performs domestic chores for a household after the occupants have retired for the night.

Specter (or Spectre): Another term for ghost.

Spirit: A discarnate being, or ghost, that exists in an invisible realm.

Spirit Photography: A term used for both legitimate attempts to capture ghosts and paranormal energy on film. Also, for the work of fraudulent photographers during the Spiritualist era.

Spirit Rescue: Attempting contact with entities, intended to alleviate the entities' distress and aid them in the resolution of their conflicts, and in "crossing over" to a higher, spiritual plane.

Spiritualism: A faith based on the idea that life continues after death and that communication between the living and the dead can, and does, take place. Spunkies: The sad spirits of unnamed, unchristian or unbaptized children, believed by old Gaelic and English tradition to wander country roads in search of someone who will name them.

Succubus: "Female" counterpart of the incubus, a demonic entity said to inspire lust in men (and most inconveniently!), sometimes capable of physically attacking and inflicting injuries (bruises & slashes). Following a nocturnal visitation from a succubus, the human victim will always feel ill and depleted of vitality, and inexplicably "un-clean."

Supernatural: Events or

happenings that take place in violation of the laws of nature, usually associated with ghosts and hauntings.

afterlife Almost every society and religion has some belief in life after death. The Christian version of the afterlife involves some type of judgment upon the body's demise and the assignment of the soul to either Heaven or Hell, while many societies believe that life simply continues in another plane of existence.

Synchronicity: Unexplained system of causal interaction which binds together events, actions and thought, manifesting as uncanny coincidences. Term for and existence of this phenomenon was first proposed by pioneering psycho-analyst, Carl Gustav Jung (a contemporary of Sigmund Freud). Synchronicity indicates there is more to the Universe than our understanding of simple cause and effect, and that the subtleties of the mind and matter are somehow interconnected.

Table-tipping: An experiment in psychokinesis (PK) which can fairly easily be replicated. Three or four participants lightly place their fingers along the edges of a small table top, then in unison chant "table move, table move..." With sufficient cooperation and concentration, and after several minutes of chanting, the table should start to wobble, pivot on its legs and possibly even lead the participants on a scurry about the room.

Talisman: A design or inscription that is worn, carried or displayed, for the purpose of invoking strength, power, protection or the aid of spirits.

Telekinesis: A psychic phenomenon where-in objects are remotely displaced and moved around, solely by the powers of the mind.

Teleportation: The appearance, disappearance, or movement of human bodies and physical objects through closed doors or over some amount of distance using paranormal means. Such events often are reported to take place during hauntings.

Thought Transference: The telepathic transmitting of images and messages from the mind of one person to that of another.

Ultra-terrestrials: Beings who appear human and visit our

plane of existence with some form of message or mission, then inexplicably vanish. Speculation abounds!

Vampire: A demonic (?) entity in the form of a deceased person, which perpetuates itself by draining the blood or psychic energy of the living.

Vortex: pl. Vortexes or Vortices. An anomaly which sometimes shows up in still photographs taken at the site of a suspected haunting, appearing as a translucent white, tube or funnel shaped mass. Some researchers believe this may be a porthole to the spirit realm.

Wraith/Wrayth: The image of a person appearing shortly before or after his or her death; term can also be applied to a ghost. Also, an apparition that is generally supposed to be an omen of death.

Zarcanor – A malevolent spirit which attacks people while they're asleep, inspiring nightmares, and sometimes even inflicting minor injuries such as scratches, bruises and what appear to be finger marks. The name is possibly of Slavic origin.

Zoomorphism: Representation of a deity or devil with animal attributes.

TO ALL MY SUPPORTER AND FRIENDS WITHOUT YOUR SUPPORT THIS WOULDN'T HAPPEN.

THANK YOU TO NEW ENGLAND GHOST FOR MY TRAINING AND EDUCATION.

If you believe you are being Haunted please contact us at ,

hauntedinnewengland@gmail.com or through Facebook

Or to find a Paranormal investigator @

http://www.paranormalsocieties.com/

Other books by Eric Perry

The Haunting of Ransom O.Gore

The Haunting of High Rock Tower

Paranormal Question of the Day Book 1 2014

Paranormal Question of the Day Book 2 2020 Cryptid edition

Bigfoot , Myth or Legend

Real New England Haunts two

Farms in New York , Picture book

True New England Haunting's in New England 2013

videos by Haunted in New England on Amazon

Haunted in New England, The Governor's Mansion

Haunted in New England, Historic Stone Cottage Lynn Ma

Haunted in New England ......Haunting of Aggie

Haunted Productions Presents Haunted in New England, Haunting of Dolly

Haunted in New England Presents The Haunted Victorian Mansion

www.ingramcontent.com/pod-product-compliance
Lightning Source LLC
Chambersburg PA
CBHW051915210526
45473CB00006B/2026